Cosmic

The Universe File

D1371703

Discovery Channel School
Science Collections

DISCOVERY CHANNEL SCHOOL

1 2 3 4 5 6 7 8 9 10 PO 06 05 04 03 02 00

Discovery Communications, Inc., produces high-quality television programming,
interactive media, books, films, and consumer products. Discovery Networks, a division of Discovery
Communications, Inc., operates and manages Discovery Channel, TLC, Animal Planet, Discovery Health Channel, and Travel Channel.

Writers: Jackie Ball, Sara Heiberger, Katie King, Kimberly King, Lisa Krause, Monique Peterson, Gary Raham, Catherine Valeriote. **Editor**: Katie King.
Photographs: Cover, Keck observatory, ©Roger Ressmeyer/CORBIS; p. 3, Maya calendar, Antochiw Collection, Mexico/ET Archive, London/Superstock; pp. 4–5, space,
©John R. Foster/Photo Researchers, Inc.; p. 6, planet, ©PhotoDisc; p. 8, Milky Way, NASA; Eta Carinae, NASA; Pleiades, NASA; p. 9, supernova, NASA; sombrero galaxy,
©PhotoDisc; Horsehead nebula, NASA; p. 15, planets, ©PhotoDisc; p. 16, Atwood celestial sphere, courtesy Alder Planetarium & Astronomy Museum, Chicago, IL; Zeiss
Mark IX, ©American Museum of Natural History; Armillary sphere, ©Gianni Tortoli/Photo Researchers, Inc.; sextant, ©PhotoDisc; p. 19, Star of Joy, *Hokule'a,* ©Painet; pp.
18–19, Robin Lee Graham, Bettmann/CORBIS; p. 20, Annie Jump Cannon, Harvard College Observatory/Science Photo Library; spectroscope, ©Charles D. Winters/Photo
Researchers, Inc.; p. 21, Jocelyn Bell Burnell, ©Jonathan Blair/CORBIS; crab nebula, ©PhotoDisc; p. 24, Maya calendar, Antochiw Collection, Mexico/ET Archive,
London/Superstock; p. 25, Stonehenge, ©Archive Photos; Big Horn Medicine Wheel, ©John A. Eddy/VU; p. 24–25, pyramid, Superstock; p. 28, Los Angeles in 1908 (IDA),
courtesy International Dark-Sky:www.darksky.org; Los Angeles in 1988 (IDA), courtesy International Dark-Sky:www.darksky.org; p. 29, Hale Bopp, NASA; all other photos
©Corel. **Illustrations**: pp. 12–13, "The Lives of the Stars," by Christopher Burke; pp. 22–23, time machine, by Dave Jonason. **Acknowledgments**: p. 18, "Star of Joy," excerpts
from VOYAGE OF REDISCOVERY: A CULTURAL ODYSSEY THROUGH POLYNESIA, by Ben R. Finney. University of California Press, 1995. Reprinted with permission; p. 19,
"Around the World Alone," excerpts from DOVE, by Robin Lee Graham. HarperCollins, 1991.

Cosmic

To Infinity and Beyond

Look up at the sky on a clear, moonless night. It's an awesome spectacle. Early in history, our ancestors gazed up at this same sight, but they had very different ideas about what they saw. They saw an infinite, timeless universe with the Earth as center stage. They used the stars to track the seasons, navigate, and even exchange tales about the pictures they saw in the night sky.

Today we know that we are not the center of the universe. The space is so vast that it can seem unknowable, and most of it is too far away for us to see. Nevertheless, we've made much progress: Scientists believe they can explain how the universe began, and how it's changing.

In COSMIC, Discovery Channel takes a closer look at stars, celestial objects, and galaxies to gain a better understanding of what makes up the known universe.

The Universe Files

How does this Maya carving mark the passage of time? Find out on page 24.

Final Project

Universe

Early on people gazed up at the night sky in wonder. They may have been thinking the same thing we are thinking today: What is the connection between the brilliant display up there and us, down here on Earth? How are we part of the same immense world? And how did that world begin?

Over a very long time and a great deal of thought, scientists developed a theory. Today many scientists think that there was a time before time when all matter, all energy, all space was condensed into something the size of a pinpoint. This was our universe in the moment before its birth. Then a huge explosion occurred, one that produced nothing but hot energy. In the first few minutes the universe expanded to 2,000 times the size of our Sun. It became a hot swirling mass, as dense as iron. Not even light could shine through it.

About 300,000 years later, when things cooled, atoms formed. The elements hydrogen and helium were created: Gravity brought them together in clumps. These clumps became the seeds of galaxies. Stars formed, including the Sun, our closest star. It took a long time for the universe to form. Scientists refer to the formation of the universe as the Big Bang theory. Other theories have also been offered, but the Big Bang has the widest acceptance.

There were no eyewitnesses, so it's impossible to be absolutely sure what really happened—and we probably will never know. It's unlikely that the Big Bang theory, or any other theory, will ever be proved as fact. However, we do know that we are only a small part of an incredibly big hierarchy, and one that is clearly organized. Our planets revolve around the Sun, and our solar system moves along the spiral arms of our galaxy, the Milky Way. Gravity holds together this galaxy—made up of billions of stars, dust, and gas. The Milky Way is just one of the millions of galaxies scattered across the universe. In between galaxies are enormous voids, or spaces full of nothing.

Scientists know that the universe is continually expanding, but it's hard to know what that means. Some scientists theorize that gravity will cause the universe to collapse, while others think it will expand forever. Since the nature of science is to continually re-examine theories based on new information, entirely new theories may arise. Meanwhile, we continue to question our own existence and examine the mysteries of the universe.

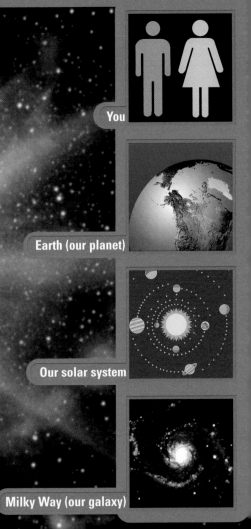

You

Earth (our planet)

Our solar system

Milky Way (our galaxy)

Activity

OUR EXPANDING UNIVERSE Create a model of the universe to show how it is expanding. For materials, use balloon, string, ruler, pen, and paper. Inflate a balloon to cm in diameter. Place 6 dots on the balloon: Label one d home (the Milky Way Galaxy) and the others A–E. The ot dots represent galaxies formed in the early universe. Measure the distance from home to each dot and record on a chart. Continue to inflate the balloon in 5-centimete increments and measure and record for the remaining galaxies. How does the distance from home to the other galaxies change when you inflate the balloon? Which do moved the greatest distance?

Meet Big Red

If you could talk to a big star . . .

Q: Hello? Anyone out there?

A: No need to scream. I can't hear you anyway. Sound doesn't travel in space.

Q: Then how can I conduct this interview?

A: Well, use your imagination! If you could talk to me, what would you ask?

Q: Well, what's up?

A: I'm up. Way up . . . in the universe. I don't think we've formally met. I'm a red giant. But you can call me Red. All the other stars do.

Q: How many other stars are you talking about?

A: In which galaxy?

Q: Pick one.

A: Okay, I'll pick the one Earth is in—the Milky Way. There are about 200 billion stars in the Milky Way galaxy alone. Of course, there are billions more stars in millions of other galaxies out there.

Q: And are they all like you? Red and big?

A: Nope. Every star is different. White dwarfs, stars like the Sun, and red giants, like yours truly, are really quite different. We stars change as we get older.

Q: You mean, you get wrinkles and gray hair and stuff?

A: Um, not quite. If I could take you on a star-studded tour, you'd understand where I'm coming from—and where I'm headed.

Q: Okay, let's go.

A: We can't. You'd never get there. The star closest to Earth, the Sun, is 93 million miles away. Even if you traveled at jet speed (600 miles per hour), it would take you over 17 years to get there. And you'd never make it because you'd burn up.

Q: Whoa! If stars are so far away, why can I count hundreds of them in the sky at night?

A: Because they give off energy and light: They shine. They shine brightly enough to see them—though they're millions and millions of miles away.

Q: How do stars get that brilliant shine? Do they use makeup like Hollywood stars?

A: They don't use makeup—but it's their makeup that makes them shine. It all starts when stars form in a cloud of dust and gas. That cloud starts collapsing under its own gravity. As the cloud clump continues to collapse, it grows hotter and hotter. When a clump's temperature reaches 18 million degrees Fahrenheit (10 million degrees Celsius), nuclear fusion reactions start. And voilà! A star is born!

Q: And how does it shine?

A: Now we're getting to the *core* of the matter: the star's core. Inside the core, hydrogen gas is converted to helium gas. That's what keeps the star shining.

Q: Isn't that what happens in the Sun's core, too?

A: Of course. Remember, the Sun is a star, too. Inside the Sun's core, hydrogen is changed to helium. That releases tons and tons of energy, which makes its way into the universe as light and heat.

Q: So, are all stars suns?

A: Definitely not. There are many stars that have more mass than the Sun and they are quite different. Supergiants and neutron stars are just two examples.

Q: I've heard the Sun is almost 5 billion years old. Are other stars that old, too?

A: Absolutely. And as stars get older, they change. Not that you humans can ever see that change. But I've been around for billions of years—from the time I was a baby star to the old red giant I am today. So I've seen thousands of stars come and go in my day.

Q: So, what will happen to our Sun?

A: Well, you know someday it will become a red giant, like me.

Q: Wow. So, how old are you?

A: About 10 billion years old. I know, I know—I don't look it.

Q: Can you tell me about some of the highlights of your life?

A: Well, billions of years ago, my hydrogen ran out, my core collapsed, and my atmosphere expanded and cooled, turning me into a red giant.

Q: And is that the end of the story?

A: Oh, no—there's a lot more to come. Eventually gravity will force my core to collapse and I'll become a planetary nebula. And then—much later on—only my core will be left: I'll become much smaller, and they'll call me a white dwarf. But that's a long way off.

Q: So why are you called a red giant? Why not a green giant? Or a purple one?

A: It's about heat. A star's color tells you its temperature. Red stars are the coolest, if I do say so myself. Blue or blue-white stars are hot stuff—they have the highest temperatures. And it's a good thing, too, because without our various temperatures, the universe would be another type of place entirely.

Q: What do stars' temperatures have to do with the universe?

A: A lot. There'd be no planets, for one thing. Scientists think very young stars that are surrounded by heat and dust may be on the verge of condensing into planets. They also think that lots of stars were born with an entire planetary family.

Q: So that means there may be undiscovered planets out there?

A: You're almost as bright as I am! There could be millions of other planets out there.

Q: How can we find out?

A: Come back and chat with me in another few billion years—if you can recognize me! It just may take that long to discover everything out here.

Activity

SEEING STARS You know that the Sun is a star, but is every star a sun? Go to your local library or the Internet and do some research. Choose one star and compare it to the Sun. Then draw a Venn diagram to show the similarities and differences between the two.

Spaced-out Spectacles

When you look up at the night sky, much of what you see is pretty ancient "stuff." Although objects look very close together, they're actually vast distances apart! Here we see galaxies, stars and star groupings, and nebulae. But many more objects exist in our universe, such as planets, satellites, comets, gravitational fields, and forms of radiation.

Milky Way Because Earth lies inside the spiral arms of the Milky Way, the few hundred billion stars of our home galaxy appear as a bright band of stars across the night sky. Dark splotches among the stars are massive globs of interstellar dust and gas that block the light from stars behind them. The nearest star to us in the Milky Way is the Sun, a medium-size star.

Pleiades Star Cluster Glittering like jewels in the night sky, the Pleiades (PLEE-uh-deez) is a cluster of new stars over 400 light-years from Earth. Though only a few are visible to us, the cluster contains hundreds of stars. The bluish glow that seems to surround the stars is a cloud of interstellar gas and dust, which scatters blue light from the stars. Different people have called the cluster "Seven Sisters," "Bunch of Grapes," and "Sailing Stars."

Eta Carinae When supermassive star Eta Carinae (ay-tuh cuh-REE-nay) exploded about 150 years ago, it became the brightest object in the sky. Eta Carinae released as much light as a supernova, but it managed to survive as a star. Today the debris stretches over a distance equivalent to the diameter of our solar system. Its outer edges are moving away from the center at about 1.5 million miles per hour.

Sombrero Galaxy
Viewed from the side, this spiral galaxy bulges in the middle like a sombrero, or Mexican hat. Scientists suspect that a black hole as massive as 1 billion Suns may lie at the galaxy's center. Around its center lie billions of old, faint stars that form the enormous bulge of light.

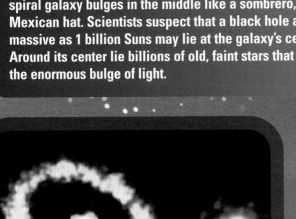

Supernova Remnant
Rings of glowing gas are all that remain of a supernova explosion. While this star ended its life in a spectacular eruption of energy and material more than 170,000 years ago, light from the catastrophe didn't reach Earth until 1987. For a time it burned bright in the sky, then faded to the glowing rings you see in this image.

Horsehead Nebula
The "horsehead" in this photo is a dense cloud of interstellar dust and gas. Found throughout the universe, gas and dust fill the spaces between the stars. In some places they gather in great enough quantities to form a spectacular cloud, called a nebula. Dust in the horse's "head" is so thick that it blocks the light behind it. Dust and gas also make for the brilliant red color in the nebula. The blue color comes from dust that reflects the light of nearby stars.

Mapping the Night Sky

North America (between 40° and 50° latitude), early March, 10:30 p.m.

What do you see when you look up at the stars—a bear? A scorpion? Ancient astronomers identified these and other patterns in the night sky. People told myths and stories about the stars, and used myths to navigate ships at sea.

Today we call these star patterns constellations, but we know now that they aren't true groups of stars. Stars in the same constellation may lie many thousands of light-years apart. Astronomers divide the sky into 88 official constellations. Most stars are in regular and predictable motion, so depending on the time of year and which hemisphere you are in, you'll see different constellations. This map shows what constellations or star groups you'd see on a clear, moonless night in early March, if you were stargazing somewhere in Canada or the northern United States. Look closely— what you think is a star may actually be a galaxy!

Polaris

Also called the Pole Star and the North Star, Polaris sits almost exactly over the North Pole. Polaris has long been important to navigators because of its position. It is located at the tip of the handle of the Little Dipper (part of Ursa Minor).

Big Dipper

The Big Dipper—a star group—appears here in the sky in early March, 10:30 p.m., but by 3:30 a.m. it's moved far to the west. What's happened? The stars don't really move, of course, but the earth is constantly on the go. Imagine the sky as the inside of a great hollow sphere, with the stars as fixed points on the sphere. Picture the earth as the center of that sphere.

Where you are on the earth and how the earth is moving determines which parts of the inside of the sphere you see.

As the Earth revolves on its axis, the stars and constellations appear to revolve in the sky around a point directly over the axis. The earth not only revolves on its axis, but it also changes position as it orbits the Sun.

So in late September the Big Dipper will appear low in the sky. And in March of next year, it will appear to be in the same place it was this March.

HERCULES

CORONA BOREALIS

SERPENS CAPUT

Arcturus

VIRGO

Spica

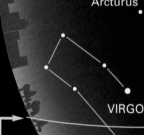

The Moon's path across the night sky

Astro-WHAT?

All of the planets in our solar system orbit the Sun in a band of the sky known as the ecliptic. Ancient astronomers divided the stars in the ecliptic into 12 constellations. Most of these constellations are named for animals or mythological creatures, so Greeks gave this band of stars the name *zodiac*, meaning "circle of animals." The zodiac is used in astrology, the belief that the planets' positions can influence the future. Astrology is not the same thing as astronomy, the scientific study of the universe. Astrology has no scientific basis at all.

North

ANDROMEDA GALAXY
The closest major galaxy to Earth, the spiral-shaped Andromeda Galaxy is the most distant object visible to the naked eye. It lies about 2.5 million light-years from Earth and contains over 200 billion stars.

BETELGEUSE
As a red supergiant, Betelgeuse boasts a distinctive orange color that stands out against the mostly blue stars of Orion. Its brightness varies over a period of about seven years and is unpredictable, sometimes changing in just a few weeks.

STAR ATTRACTION: ORION
With its distinctive "belt" of three bright stars, Orion is one of the most easily recognized constellations. The Greeks saw the star group as a hunter, but Native Americans saw a group of running deer. Syrian astronomers believed the stars formed a giant called Al Jabbar. In the northern hemisphere, you cannot see Orion in June. However, you can see it in Chile in June.

CEPHEUS

ANDROMEDA

URSA MINOR
Little Dipper

Polaris

PERSEUS

CAMELOPARDALIS

Capella

AURIGA

TAURUS
Aldebaran

Big Dipper

URSA MAJOR

Linx

CANES VENATICI

Castor

Pollux   GEMINI

Betelgeuse

LEO MINOR

CANCER

ORION

LEO

Procyon

CANIS MINOR

Regulus

Activity

Sirius

CRATER   Alphard

CANIS MAJOR

HYDRA

PUPPIS

STAR BRIGHT: SIRIUS (found in Canis Major)
While Sirius is not the brightest star in the universe, it appears that way because it is so close to Earth—just 8.6 light-years away. It shines as bright as 23 Suns. Because Sirius lies in Canis Major (the Great Dog), it is sometimes called the Dog Star. Its faint companion, Sirius B, is a white dwarf nicknamed "the Pup."

SIZING UP THE STARS To surfers, a closed fist with thumb and pinky extended means "hang ten," but astronomers use this as a way to measure the sky. The circle of sky is divided into 360 degrees. The "hang ten" sign shows about a 20-degree distance when holding your hand out at arm's length. This is roughly the length of the Big Dipper. A closed fist measures about 10 degrees, and one finger equals one degree. At night, identify a major constellation, like Orion or Ursa Major. How many degrees wide are these constellations? How many degrees is it to the next big constellation? Make some measurements, then check your notations using a detailed star chart. How accurately did you measure the stars?

AND NOW...THE LIVES OF THE STARS!

Path of the Sun
(And other stars as massive as the Sun)

MAIN PHASE	RED GIANT
10 billion years	2 billion years

For most of its life, the Sun (and other stars with a mass like the Sun) shines steadily. At its core, nuclear reactions are burning hydrogen and converting it to helium. The Sun takes about 10 billion years to use up the hydrogen in its core: It is now about halfway through.

When the star's hydrogen runs out, gravity causes its core to shrink and get hotter. The cooler outer layers expand and glow red. The star becomes a red giant. It now burns helium at its core. When this runs out, gravity once again causes the core to collapse.

Path of Massive Stars
(Stars 10–50 times as massive as the Sun)

MAIN PHASE	RED SUPERGIANT	SUPERNOVA
1–20 million years	2–5 million years	Immediate; glows brightly for several months

Stars more massive than the Sun live fast and die young. They burn through their hydrogen fuel in less than 20 million years. Initially they are blue and shine much hotter and brighter than the Sun and other sun-like stars as they race toward the end of their lives.

As a massive star runs out of hydrogen, it begins to expand. Its core heats up while the outer layers cool and glow red. At its core, the red supergiant burns its helium to make other elements, including carbon and oxygen. Eventually it begins to make iron, the heaviest of elements.

Iron absorbs energy rather than releasing it. In a split second, the star collapses under its own mass, condensing its core into a dense nucleus. The nucleus reacts to the squeezing by releasing massive shock waves, exploding the star into a supernova. A supernova can shine brighter than 1 billion Suns.

The Sun is the largest body in the solar system, but it is also a star—a pretty average star. Out in the universe the Sun is one of millions of stars. Like all stars, the Sun went through many complex changes after it was born. This took over hundreds of thousands of years. Once a star is formed, its mass will determine how long it shines and what happens during the rest of its life. For example we know that our Sun will become a red giant in 5 billion years. But there's more to come . . . it's got a long life ahead!

The timeline below shows several paths a star can take. The first is the Sun's path. (Stars with the same mass as the Sun also follow this path.) The second path is for stars more massive than the Sun. Of course, our Sun will eventually burn up and disappear from view, like many stars. But don't worry—that's a long way away!

PLANETARY NEBULA
A few thousand years

Heat from the collapsing core is transferred to the red giant's outer layers, causing a final reaction that blasts off the star's outer layers. A shell of material drifts outward into space, leaving a glowing core. Light from the core often illuminates the cast-off material, which is a planetary nebula. Early astronomers thought they looked like disks of planets.

WHITE DWARF
Fades slowly

Only the core of the original star remains. It glows for a long time as a white dwarf, shining weakly. When it loses the last of its heat, it becomes a black dwarf and disappears from view.

NEUTRON STAR
Stars 10–30 times as massive as the Sun

The core that remains becomes a star made of neutrons, which are extraordinarily dense. A teaspoon full of neutron star material weighs about 10 billion pounds, roughly the weight of the Sears Tower in Chicago, Illinois.

BLACK HOLES
Stars 30–50 times as massive as the Sun

Sometimes the core that remains after a supernova is so massive that nothing can support it against its own gravity. It collapses into itself and becomes a black hole. Gravity is so strong in a black hole that nothing—not even light—can escape.

Activity

GRAVITY WELLS Gravity affects how stars are formed. This demonstration shows you how star formation works.

Materials:
- 2 shallow bowls (of different sizes)
- bag of sand
- newspaper

Spread out the newspaper and place the two bowls on it. Pour out the sand in a wide swath, so that it falls into and around each of the bowls.

1. What happens to the sand in the bowls?
2. What do you think the sand outside the bowls represents?
3. How does the size of the bowl affect the outcome?

Check out the answers on page 32.

"Put three grains of sand inside a vast cathedral, and the cathedral will be more closely packed with sand than space is with stars."
— Sir James Jeans, English astronomer (1877–1946)

The Outer LIMITS

A Journey of Tens

To get an idea of the size of the universe, you have to think big. One way is to use orders of magnitude. Beginning with 1 meter, the graph below shows the relative size of things in the universe as you increase the distance scale exponentially by powers of 10. (See Too Many Zeros, opposite page.)

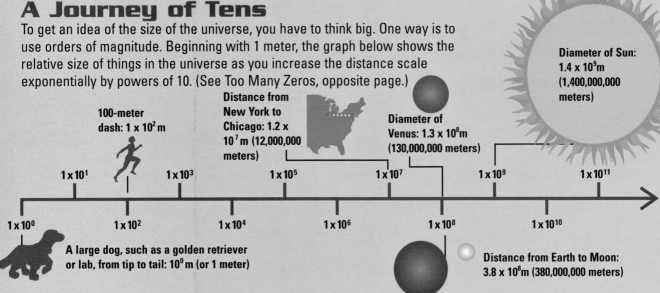

Diameter of Sun: 1.4×10^9 m (1,400,000,000 meters)

100-meter dash: 1×10^2 m

Distance from New York to Chicago: 1.2×10^7 m (12,000,000 meters)

Diameter of Venus: 1.3×10^8 m (130,000,000 meters)

1×10^1 1×10^3 1×10^5 1×10^7 1×10^9 1×10^{11}

1×10^0 1×10^2 1×10^4 1×10^6 1×10^8 1×10^{10}

A large dog, such as a golden retriever or lab, from tip to tail: 10^0 m (or 1 meter)

Distance from Earth to Moon: 3.8×10^8 m (380,000,000 meters)

Jet Speed or Light Speed?

Traveling in an airplane is a fast way to go, but it won't work in space. The speediest way would be on a beam of light. It zips through space at 186,000 miles per second. At the speed of light, you could circle Earth seven times in one second, and travel to the Moon and back in less than three seconds. In a year, a beam of light travels about 5.88 trillion miles. Astronomers use light-years to describe some of the tremendous distances to places beyond our solar system. Below are some travel times from Earth to other popular places in the universe, either in a jet or on a light beam.

Destination	Jet Speed (600 miles/hour)	Light Speed (186,000 miles/second)
Sun	17 years, 18 months	8.5 minutes
Mars	8 years, 10 months	5 minutes
Pluto	690 years, 1 month	5 hours, 25 minutes
Alpha Centauri (closest star)	4.8 million years	4.2 years
Sirius (brightest star as seen from the earth)	9.6 million years	8.4 years
Pleiades Star Cluster	450 million years	400 years
Center of the Milky Way	42 billion years	38,000 years
Andromeda Galaxy	2.5 trillion years	2.2 million years

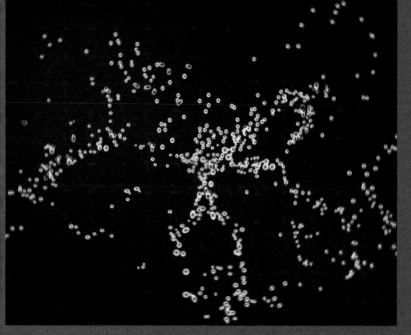

Stick Man in the Sky

This isn't a doodle of a stick man—it's a map of galaxies! In 1986, astronomer Margaret Geller led a team of researchers to pinpoint the location of more than 1,000 nearby galaxies. The map shows that galaxies are not evenly spaced in the universe, but grouped together in narrow bands with large spaces or voids between them. Geller nicknamed the image the "stick man" because it resembled a stick figure with arms outstretched.

Too Many Zeros

Scientists save space with very large and very small numbers by using scientific notation, or "powers of ten." For example, instead of writing out 150,000,000,000 meters (the distance from the earth to the Sun), mathematical shorthand makes the number more manageable: 1.5×10^{11} m. The powers of ten notation refers to the magnitude of the number. The "power" indicates how many times 10 is multiplied by itself.

Here are some more examples of how this works:

Number	Number of "10s"	Final Notation
500,000	$5.0 \times 10 \times 10 \times 10 \times 10 \times 10$	5×10^5
2,500	$2.5 \times 10 \times 10 \times 10$	2.5×10^3
300	$3.0 \times 10 \times 10$	3×10^2
10	10	1×10^1

Your Place In Space

Just in case you ever need to give your cosmic address to an interstellar pen pal, here's your place in space:

Shelley Starfinder
4 Mars Lane
Jupiter, Florida
United States
North America
Planet: Earth
Solar System: Milky Way Galaxy
Local Group: Virgo Cluster
Local Supercluster
Universe

US 10¢ Skylab

Activity

POWERS OF TEN Use scientific notation to convert the following distances:

1. Distance from Earth to Saturn: 1,200,000,000,000 meters
2. One light-year: 9,500,000,000,000,000 meters
3. Distance to nearest star: 40,000,000,000,000,000 meters
4. Diameter of Milky Way Galaxy: 770,000,000,000,000,000,000 meters
5. Distance to Andromeda Galaxy: 2,100,000,000,000,000,000,000 meters
6. Distance to most distant galaxy: 140,000,000,000,000,000,000,000,000 meters

Answers on page 32

STARS.
of the Show

Modern planetariums may be high tech today, but globes and other models representing the stars have been around a long time.

Rotating Rooms

Stars have always been important for telling time, tracking the seasons, and finding locations and directions. But people needed ways to study them. It was difficult to map something dome-shaped onto a flat surface, and so a sphere was created to map the night sky.

People used globes as early as 2,000 years ago. But until the 17th century, globes were always solid with the stars placed on the outside. This meant the stars were shown in reverse, and studied from a point of reference somewhere outside the universe.

In 1664 Andreas Busch built the Gottorp globe to represent the heavens from inside a dome. The first step toward a modern planetarium, this globe resembled a hollow sphere big enough to hold 10 people inside. The constellations were painted on the inside. It took 20 years to make and was only 10 feet in diameter. To show the motion of the stars across the sky, the 3.5-ton globe was water-powered to rotate once every 24 hours.

In 1913 the Atwood globe, built at the Chicago Academy of Sciences, was lit from the outside by electricity and could hold 17 people. Like the Gottorp globe, this globe also rotated.

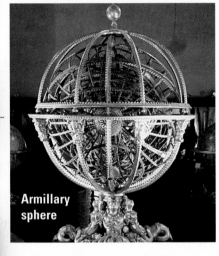

Armillary sphere

The Whole World in Your Hand

Could you understand the stars and planets by using a device small enough to fit on your desk? Before telescopes, globes, and modern planetariums, astronomers and mathematicians wanted tools to help them figure out celestial objects' locations and predict their motions. Luckily, the ancient Greeks invented the armillary sphere, an object that demonstrated the motions of celestial objects through the sky. (This predated the idea that the earth and planets move around the Sun. Later, the design of these spheres changed so that the Sun—and not the earth—was positioned in the center.) People still own armillary spheres today, but they're mostly used for decoration.

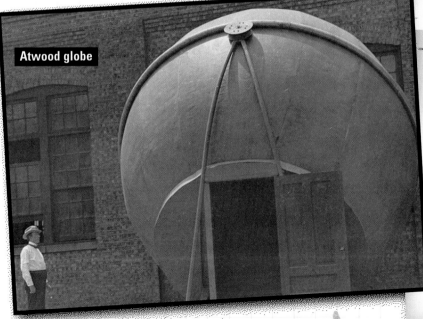

Atwood globe

Glass-Domed Spaceship?!

In the 1920s, inventors at the Carl Zeiss Optical Company found a better way to view the stars. They invented a rotating projector that casts stars on the inside of a dome. Unlike earlier models, the room itself did not have to rotate. The modern planetarium was born.

A planetarium projects images of the solar system, stars, and galaxies onto a large sphere. Observers gaze at the ceiling. Modern planetarium star projectors show over 9,000 stars, as well as the Sun, Moon, and planets. The projectors have many lenses and are a little like elaborate flashlights. The bulbs of the projectors are covered by blackened lenses dotted with tiny holes representing the stars. Only the light of these "stars" is projected on the domed ceiling.

The whole projector rotates. Turning it one way shows the rising and setting of the Sun, Moon, planets, and stars in a 24-hour day; turning it another way shows the changing positions of the stars over the course of a year, and what the sky might look like in the past or future. In a planetarium, viewers can hurtle through a day in less than two minutes!

The Virtual Reality Theater, in the new Hayden Planetarium, at the American Museum of Natural History in New York City is one of the most technologically advanced planetariums in the world. The Zeiss Mark IX, its new custom-designed projector, takes you anywhere in the universe!

Frank Summers, an astrophysicist at the museum says, "Planetariums have always shown the stars as they look from Earth, but with this digital technology we can actually move out into the stars and through local galaxies. It's like being in a glass-domed spaceship."

FUN FACTS ABOUT ZEISS MARK IX

► Zeiss uses fiber optics to show all the stars visible to the naked eye— as well as some so tiny you need binoculars to see them!

► It uses seven high-resolution digital projectors, each covering a portion of the dome.

► The images' edges overlap and are digitally mixed to blend together perfectly.

► You can see the sky from Earth and from every planet in the solar system!

The Zeiss Mark IX

Eyes on the Skies

"Star of Joy"

Before people had modern instruments like the compass, they often relied on the stars to guide them across the high seas. In 1976, a 17-man crew set off to retrace routes of ancient voyages between islands in the South Pacific. Ben Finney, a professor at the University of Hawaii, led the crew as they traveled on *Hokule'a,* meaning "star of joy," a Polynesian canoe.

Mau was their navigator who successfully led *Hokule'a* from Hawaii to Tahiti using only the stars, Sun, wind, and waves for guidance. After 34 days at sea, the crew arrived in Tahiti.

This passage is from *Hokule'a,* Finney's record of the epic voyage.

Mau usually set his course by watching the position of the stars.

The particular point on the eastern horizon toward which Mau is now aiming the canoe is where the star Tumur rises. Tumur is Satawalese [Mau's native language] for Antares, a red giant star 285 times larger than our Sun. Its reddish glow dominates Scorpius, the constellation Polynesians call the "Fish-hook of Maui," after the demigod who fished up islands out of the sea.

Using Mau's star compass, we can see that instead of blowing from the northeast, it [the wind] is coming mostly from the east-northeast. That prevents us from always pointing the canoe as high as Antares. Much of the time we are heading for the next star compass point over from Antares—the rising point of Shaula.

Twilight

Steering became particularly challenging in the early evening, when the stars were hard to see.

Antares is of course not always conveniently on the horizon when Mau wants to make a star sight. It does not rise until an hour or so after dark, and toward midnight it is already too high in the sky to give an accurate bearing. But Mau knows the sky well enough after three decades of practice to tell at a glance where Antares should cut the horizon . . .

Should the horizon be so cloudy that no horizon stars are to be seen, Mau can still get his bearings by sighting on whatever stars are visible. A glance at Polaris, the Southern Cross, or other stars and constellations is sufficient. Mau knows the shape of the sky so well that he has only to see the smallest portion to find his way.

The *Hokule'a* on its journey

AROUND THE WORLD ALONE

San Pedro, California, July 1965

Relying heavily on his sextant, sixteen-year-old Robin Lee Graham set sail aboard his 24-foot sailboat, *Dove*, for a voyage around the world. The trip took five years. He survived hurricanes, nearly lost his mast, and was swept overboard in high seas. Graham even got married along the way! Though he had modern instruments, he still relied on the predictability of the stars' motion to help him navigate. Here is an excerpt from his journal.

Robin Lee Graham aboard his sailboat

A sextant, used for navigation

Day Three

Just had a dinner of canned turkey and yams, which stuck to the roof of my mouth. Took my first moon LOP [line of position] with a sextant.

Day Ten . . .

I hit the trades, which pushed Dove along 120 miles in 24 hours, and the clear night sky allowed me to take my first star fix. I was really excited about this and taped: It's two o'clock in the morning and I know exactly where I am. That's kind of fun . . .

To the non-sailor, navigation may seem like witchcraft, but really it's not at all difficult. The sextant is key to it all. With this instrument I measure the altitude above the horizon of the sun, moon or stars, then mark the time to the second on my chronometer. After that it's simply a matter of looking up the nautical tables, making additions and subtractions which wouldn't strain an average ten-year-old and pinpointing my exact position on the charts.

Activity

GOING TO SCHOOL As you already know, the Sun is a star. Since ancient times people have used the Sun to navigate. Picture your route to school. Now pretend that there are no signs or streets to guide you there. How would you use the Sun to get to school? What natural items (like large boulders) could help you find your way? How would you know if you're late for school?

StarGAZERS

ANNIE JUMP CANNON (1863–1941)

As a young girl, Annie Jump Cannon climbed to the roof with her mother to look at the stars. By candlelight, she looked through astronomy books and tried to identify the stars. These were the first steps to her life's work, classifying nearly 400,000 stars. The results fill a 10-volume catalog known as the bible of modern astronomy. Thanks to Cannon, scientists know a lot about how stars are born, age, and die.

A spectroscope

"Oh, Be A Fine Girl!"

In 1896 Cannon went to work at the Harvard College Observatory. Her job was to study glass photographic plates of stars taken with a spectroscope, an instrument that contains a prism to separate light into a spectrum. Every star has a slightly different spectrum because each one is made up of different elements. First Cannon described the star and then put it in a category. She organized the star categories in a sequence from hottest to coolest: O, B, A, F, G, K, M, R, N, S. Here's an easy way to remember the sequence: "Oh, Be A Fine Girl, Kiss Me Right Now, Sweet!" In 1910 the International Astronomical Union adopted Cannon's system, now known as the Harvard System of Spectral Classification.

Keen eyesight and practice made Cannon a fast sorter. Between 1911 and 1915 she classified 5,000 stars a month, sometimes as many as three stars a minute. Variable stars—stars that glow brighter and dimmer at regular intervals—are difficult to classify. Cannon especially enjoyed the challenge of classifying these stars.

For years Cannon studied pictures by day and checked her work through a telescope at night. When asked how she could spend so much time looking at streaky pictures, she said, "They aren't just streaks to me. Each new spectrum is the gateway to a wonderful new world."

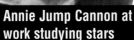
Annie Jump Cannon at work studying stars

JOCELYN BELL BURNELL (1943–)

"A ragged, scruffy piece of signal." This was Jocelyn Bell's first clue to an exciting discovery. She was the first astronomer to identify a pulsar, a new kind of star. Bell was just 24 years old.

Cambridge, England, July 1967

Little Green Men

Jocelyn Bell began working on her Ph.D. in astronomy at Cambridge University. Here she helped build a huge radio telescope. Bell and fellow students strung miles of wire antennae from 1,000 poles that were 9 feet tall, covering more than 4 acres.

In 1967 the telescope was ready to listen to the sounds in deep space. The antennae detected radio waves, and automatic pens recorded these waves as signals on long rolls of paper. Bell rode her bike out every day to oversee the operation, fill the inkwells, and analyze the charts.

The "ragged, scruffy" signal first appeared in July. Bell couldn't explain it, so she showed the results to her professor Anthony Hewish, a radio astronomer. He thought the signal was artificial, but Bell and others showed that it was coming from deep space—200 light-years away.

For a month Bell checked the telescope for another signal, but with no luck. Then one day, there it was! The pen had gone "whoop, whoop, whoop" at regular intervals over the paper. She put her ruler to the chart on the floor and saw that the radio waves were sending out "pulses" every 1⅓ seconds. The regular pulse made Bell's group wonder if the signal was coming from intelligent life. They even started calling it LGM, short for "little green men." In December Bell found another pulsing signal, but in a different part of the sky. There couldn't be two different groups of little green men signaling. The name LGM was replaced with "pulsar."

Bell concluded that these pulsars were a kind of rotating neutron star. Because they are collapsed, neutron stars are very small and dense. One pulsar might have as much mass as our Sun packed into a ball just 10 miles wide. Every time the star spins, it flashes a radio signal.

Activity

IT'S CLASSIFIED Before Cannon's system, astronomers didn't really know what they were seeing when they looked at the night sky. Nevertheless, they made detailed observations and used a brightness scale, 1 being the brightest and 5 the dimmest. Draw your own picture of stars and place the stars into 1 of the 5 classes.

TRAVEL LIGHT

Y ou are a photon of light, released when two atoms of hydrogen slammed together to make helium, and you have just escaped from the gravitational hold of a star. You are traveling 186,000 miles every second and 5.88 trillion miles in one year—the length of a light-year. This journey will last 4.6 billion light-years. You are traveling toward a planet called Earth, a loose ball of dust that is just solidifying, in a galaxy far, far away.

5.88 Trillion Miles Per Year

Destination: EARTH

Y ou travel for a half billion years while this distant planet forms from old star dust and rubble, condensing into a molten ball. Heavy elements like iron sink to the core. Lighter silicates (minerals that contain silicon and oxygen) cover the core with a mantle of partially liquified rock. Gases, including carbon dioxide, nitrogen, and water vapor, hiss out of this cauldron, releasing heat. In another 300 million years of travel, a thin crust has begun to form on this molten ball.

You travel for another—rather boring—billion years. You rarely pass close to a galaxy, much less a single star. Earth, meanwhile, is forming a harder crust and an atmosphere. Water vapor becomes trapped near the planet's surface. It evaporates, then rains, creating rivers of liquid water that pool in deep ocean basins. Earth has become a "water world" and . . . something else. A corrosive gas, oxygen, has begun to appear in the oceans and in the atmosphere. You know what this means: Simple life-forms have discovered photosynthesis—how to change sunlight into chemical energy.

You travel another 2 billion years. Earth is becoming an aquarium full of "pond scum," whose waste product, oxygen, has already "rusted" all the iron on the planet and is collecting in the nitrogen-carbon dioxide atmosphere. Now some interesting

Miles Per Second: 186,000

| 0 | 50,000 | 100,000 | 150,000 | 200,000 |

animals begin to create shells and hard parts from calcium carbonate in the water.

After another 100 million years, as you pass near a cluster of galaxies, plants and animals on Earth have discovered how to live on dry land. And 150 million years later, forests of trees cover huge areas. Terrarium Earth looks pretty healthy until suddenly, lots of volcanoes erupt, ocean currents change, a few big meteorites hit the planet—and 90 percent of Earth's life-forms die! It takes up to 10 million years for living things to completely recover.

For the next 160 million years dinosaurs—from house-size to chicken-size—inhabit the earth on land and in water. Soon some mouse-size, hairy things called mammals appear . . . But what's this? All the dinosaurs have died off. Did something suddenly change the climate? Say a meteorite or something? Once the dinos are gone, mammals start to populate the earth.

Now you are traveling near several dozen groups of galaxies, one of which contains the Milky Way Galaxy, where Earth's Sun glows. You still have 50 million years or so before your journey is over.

Finally, just 3 million years away, the Andromeda Galaxy whizzes past! The Milky Way and its neighbors, the Magellanic Clouds, are bright spots dead ahead. Earth is having another one of its ice ages and afterward some mammals called primates are playing in the trees and developing good hand-eye-brain coordination.

At 200,000 light-years from your destination, the Milky Way Galaxy slowly turns before you, an impressive pinwheel containing billions of stars and delicate, dust-cloud veils. On Earth, a new species of primate, *Homo sapiens*, looks up at the night sky and wonders about all the stars . . . but they can't yet see you.

The Milky Way stretches 70,000 light-years across, but you only need to find a small, yellow sun in one of its spiral arms. Yes, you finally see it—a pinprick of light among many others—home to the world that has been evolving as you traveled.

In all this time you have remained unchanged, in spite of all the changes taking place elsewhere.

Finally, you reach the edge of Earth's solar system. Only 11 hours separates you from your final destination. A young girl—one of those *Homo sapiens*—is eating dinner by a lake in northern Minnesota on a clear autumn night.

She looks up and sees you. Your journey is finally over.

stuff starts happening pretty quickly . . .

As you continue traveling over the next several hundred million years, life on Earth starts to get more complicated. Individual cells begin living in organized colonies. Simple animals, plants, and fungi appear in the oceans. Toward the end of this period, after several alternating cold and hot spells,

Activity

THE SPEED OF LIGHT

Referring to the Almanac on pages 14–15, find the nearest star to the Earth. Go to your local library and research the star. Trace the star's light journey to Earth. How long did it take to get here? Now make a timeline of all the major scientific advances that took place on Earth during that time.

ANCIENT SKY WATCHERS

You might think the study of the universe was a new thing, but you'd be surprised. Ancient ruins thousands of years old, all over the world, raise questions about how people have observed the night sky over the centuries. Why would people build such elaborate structures, and how are they related to the movements of the stars, planets, and our Sun?

These are the kinds of mysteries that scientists called "archaeoastronomers" try to solve. They use information from astronomy, archaeology, anthropology, and history to piece together the puzzles. They also study the ruins at different times of year, to see how the buildings might reflect changes in the sky as Earth revolves around the Sun, and as the planets appear to move across the heavens.

GIZA!

About 2.3 million blocks make up the Great Pyramid of Giza (2600 BC), which is also the tomb of Pharaoh Khufu. The pyramid's four sides face almost perfectly north, south, east, and west, a precise alignment that hints that the pyramid might have also been a sky observatory. Mysterious passageways in the pyramid might have been places for viewing the stars at certain points of the year. Perhaps ancient stargazers made their observations from these openings while the pyramid was under construction, before the passageways were sealed shut. Another theory is that the airshafts guided Khufu's soul toward his ancestors living among the stars.

The Great Pyramid of Giza at night

NEW WORLD PYRAMIDS

Throughout the jungles of Central and South America are spectacular ruins of ancient flat-topped pyramids. The Aztec, Toltec, and Maya peoples built them, starting around 1200 BC, as royal tombs and temples. To the Aztecs, the most important god and "star" was Venus. When Venus appeared in the night sky, according to one 16th-century historian, the Aztecs sacrificed captives in its honor.

Like the Egyptian pyramids, the New World temple plazas are laid out so precisely that experts believe they were for studying the night sky. The Maya also made detailed astronomical observations. In fact, their elaborate solar calendar was the most accurate in the world until our Gregorian calendar was created in 1582.

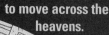

"DANCE OF THE GIANTS"

Between 4000 and 1500 BC, human laborers built thousands of stone circles and piles across Britain and western Europe, using megaliths, or massive rock pillars. Stonehenge, near Salisbury, England, is the largest of such circles (right). Its pillars weigh 4 to 6 tons; workers had to bring these to the site from hundreds of miles away. Most likely, people rolled the megaliths on logs. They also shoveled a ditch around the site, and made 56 pits. Here they may have cremated and buried the dead.

No one knows exactly how ancient peoples used Stonehenge. If you look from the center through a northeast arch toward the "heelstone," you'll see the Sun rise on the morning of the summer solstice, the longest day of the year. Perhaps this provided information to make a calendar for planting crops. Some scientists say the stones are lined up precisely enough to mark the start of spring, fall, and winter, and to predict eclipses of the Sun and Moon.

BIG HORN MEDICINE WHEEL

In North America, ancient Native Americans left behind some 50 medicine wheels, puzzling stone circles. They built the Big Horn Medicine Wheel, in Wyoming, about 600 years ago. It features 28 spokes of rocks radiating from a large pile of rocks in the center. Around the rim of the wheel are six more piles, with one much farther out than the others. A line drawn from the farthest rock pile through its center will point straight toward the sunrise of the summer solstice. Lines drawn through three other piles mark the rising points of the stars Rigel, Sirius, and Aldebaran. Native Americans may have tracked the stars so that they could predict the change of seasons. For some this may have meant that it was time to relocate. Summer solstice marks the time when tribes celebrate summer's arrival: They pray for rain and abundant crops.

Activity

SUNGAZING **Create your own observatory based on Stonehenge. Stand outside with your back against a tree or wall (your "backsight"). Look toward the setting Sun and see if it is near some point in front of you. Do not stare directly into the Sun! (That telephone pole, fence post, or rooftop will be your "foresight.") Do this for several days and see if you detect the Sun setting nearer or farther away from your "foresight." Write down your observations. Take photos periodically.**

The students in Mr. Cosmo's science class couldn't believe it. The toughest teacher in the whole school had just announced their homework was to write a science fiction story.

"Cool," Ray whispered to his friends Sonny and Celeste. "That'll be a lot more fun than those math calculations and diagrams we've been doing all semester." Mr. Cosmo was writing something on the blackboard. When he finished he turned back to the class.

"This isn't just any old science fiction," he said. "I expect you to do as much research for this assignment as you would for any class report. And it's not going to be easy. That's why I want you to divide into teams."

The bell rang and the students started gathering their notebooks and pens. "Hold up a minute," Mr. Cosmo called out. "I haven't finished explaining the assignment. Look at the blackboard: This is your story's opening sentence."

The spaceship Orion IV was launched and on its way. Its commander and crew knew their mission was dangerous but necessary. If they didn't locate new planetary systems capable of supporting life, their civilization would be doomed forever.

"Here's the science fiction part," Mr. Cosmo went on. "Orion IV can travel the speed of light. The ship is equipped with suspended animation life-support, so the crew can survive in space way beyond their average lifetimes. In other words, speed, distance, and time aren't a problem. But you have to tell me what happens next, and it's got to be based on science fact, not science fiction! Don't forget to do your research and work with each other to give me something I can believe. Oh, and try to make it exciting, too. Stories are due next week. Class dismissed."

The three friends talked over lunch. "The first thing the crew's gotta do is plan where they're going," Ray said. "Where are the best planets likely to be?" After school, they spent some time in the library gathering preliminary information. The plan was that each would sketch out a story separately and meet the next day.

"We'll take each other's ideas and check them against more research," Celeste suggested, "to make sure we got the facts right. Then we'll flesh out the best one."

Celeste's story idea

The Orion IV plots its course toward a cloud of interstellar dust and gas called the Moosehead Nebula. As the ship gets closer, its instruments pick up strange signals coming from the center of the cloud: flashes of electromagnetic radiation, pulsating at regular intervals. It's a star of some kind, sending out deadly gamma rays, and it could explode into a supernova any minute! The commander tells the crew to reverse course and activate light-speed to escape the nebula as quickly as possible. They're going to have to look for planets somewhere else.

Ray's story idea

The Orion's navigator studies the ship's charts and gets information about a star some four light-years away. It's glowing and red, and it's big—almost as big as Earth's orbit around the Sun! This star is also in the middle of a huge gas cloud called a planetary nebula. As the ship gets closer, they'll search the nebula for planets. They're pretty sure some of these planets can support life because the star is so big and bright. It's also hot—surface temperature ranges between 3,000 and 4,000 Kelvin. But does that make the planetary nebula too hot to explore?

Sonny's story idea

After scanning the galaxy and getting lots of data about different stars, the Orion plots its course for a medium-size yellow star. Although it's small compared to other stars, it's hot—the surface temperature is 6,000 Kelvin. This star has planets and other things going around it, including a big ring of asteroids 200 million miles wide. There's another ring of comets on the outer reaches—something like 100 billion of them. Can the ship make it through all this stuff in one piece? Stay tuned!

The group met in the school library the next day and exchanged their notes. "Cool ideas," Ray announced. "You guys should be in the movies."

"I dunno," said Sonny doubtfully. "Where's the story supposed to start, anyway? Mr. Cosmo didn't really say… "

"That's pretty obvious," Celeste answered. "Anyway, the whole point is getting the facts right." She took Ray's notes and wandered over to the reference section. "Now…where can I find out about a planetary nebula?" she mumbled to herself.

Ray took Sonny's idea, and Sonny took Celeste's. They got together for lunch and compared the results of their research. "Well," said Ray after they went through their notes, "it's a good thing there are three of us on this team, because only one of these ideas is going to fly."

Which story idea best accomplishes Mr. Cosmo's assignment? Why won't the other two work? Study the notes to help solve the mystery.

Use these clues...

Celeste's notes on Ray's idea

Planetary nebula—ring of gas around a red giant; part of star's outer core that's been blasted away from center

Red giant—phase of star after it has used up its fuel. Gravity takes over; core outer layers cool and expand as core collapses. Sun will reach red giant phase in 5 billion years—hot enough to boil up the oceans!

Ray's notes on Sonny's idea

Small yellow star— same type as Sun

Asteroids and comets—bad news for life on planets. See theory that six-mile-wide asteroid killed the dinosaurs. But gravitational force from star and other big planets (like Jupiter in our solar system) can keep them in place.

Sonny's notes on Celeste's idea

Nebula—cloud of dust and gas, where stars and planets form. Also what's left after a star explodes (supernova). No center of gravity to support orbits.

Neutron star—very heavy remains of star's core after supernova

Pulsar—type of neutron star that spins rapidly, sends out regular pulses of energy and gamma radiation

Answer on page 32

BLINDED BY THE LIGHT

For those who love to gaze at the night sky, the International Dark-Sky Association (IDA) is really making a difference. Founded in 1988 by astronomer Dr. David Crawford and fellow stargazers, the IDA's mission has been to preserve humanity's view of the universe.

On a clear night, you should be able to see about 1,500 stars and recognize some of the 88 named constellations. In the summer, you also might see part of our Milky Way galaxy. But even if you live in a rural area, you might only see a few stars because there is too much light. Outdoor lighting fixtures send too much light up into the sky, where it is scattered by dust and moisture in the atmosphere, creating "sky glow." Instead of seeing constellations, city observers see only a dull grey sky with a few bright stars. At the same time, astronomers are finding it harder to work.

Astronomers were among the first to recognize the need to protect the view of the night skies. With their large ground-based telescopes, astronomers search the night sky for clues about distant galaxies. The light from stars has traveled for billions of years and only a faint amount reaches Earth. Astronomers' telescopes must be located far from sources of light pollution, and they are very protective of their light-free observatories. At the Mauna Kea observatories on the island of Hawaii, the road that leads to the mountaintop (14,000 feet) is not only winding, steep, and fog-covered, it's also unlit! Night driving is strongly discouraged, and hiking is not an option.

Los Angeles, California, at night

A Good Idea

The IDA has more than 4,000 concerned members who are dedicated to educating people about how to control light pollution and the value of using efficient nighttime lighting. The IDA's task is not an easy one: People demand a safe, secure nighttime environment. But sometimes safety can mean too much light. In many places, building lights and parking lot lights are left on all night, even after everyone has gone home.

Comet Hale-Bopp

In 1996–97, news casts everywhere reported on Comet Hale-Bopp, which won't return until the year 4300. Everyone wanted a glimpse at the comet and its brilliant tail. But people in urban areas couldn't see the famous comet. Outraged, they called university astronomy departments demanding to know why. Sky glow, of course, was the culprit.

Even animals are affected by too much light. The IDA, along with environmental groups and the Florida Department of Environmental Protection, have documented how light pollution may have caused the deaths of endangered sea turtles. Hatchlings, born in the dunes on beaches, must quickly scramble to the sea. But in populated areas, hatchlings are disoriented by the bright artificial lights and are lured inland. Many are eaten by predators, die from dehydration, or are run over by cars. Currently, the IDA and environmental groups are educating people about shielded lighting practices that won't affect the sea turtles' nesting habitat. Working with the Fatal Light Awareness Program (FLAP), the IDA convinces people in office complexes to turn their lights out at night because migrating birds often collide with brightly lit buildings.

passed legislation to restrict state-funded outdoor lighting. These states joined Arizona, Connecticut, and Maine in the fight to protect the beauty of the night sky.

For visitors to national parks, the night sky is a major attraction. In Shenandoah National Park, there are now public programs on light pollution and astronomy, and they are redesigning their lighting facilities to be more environmentally sound.

Elizabeth Alvarez of the IDA sums it up: "We're here to help find solutions that are win-win, that help people put light where it's needed, when it's needed, and in the amount it's needed." Thanks to the IDA, the majestic view of thousands of stars twinkling overhead will not be lost. See for yourself: www.darksky.org

Starry Nights

According to Bob Gent, public relations coordinator for the IDA, the association's hard work is starting to pay off. "Last year, 1999, was a remarkably good year. We received a lot of great press coverage, and our Web site received thousands of hits per day from interested individuals."

The governor of New Mexico signed the Night Sky Protection Act, which outlines a new outdoor-lighting program for the state and legally recognizes the night skies as an important resource to protect. Texas also

Activity

COUNT THE STARS Cut out a square of paper 8 inches x 8 inches, and then cut a square opening in the middle. Take a piece of string and tape it to one corner of the paper and the other end to your shoulder. Look through the frame at the stars. This divides the sky into 40 pieces. Hold the frame steady and count how many stars you see inside it. Do this 5 times in different parts of the sky. When you're done, write down the number of stars you see. Divide this number by 5 and multiply that answer by 40. You have estimated the number of stars you can see.
(Source: International Dark-Sky Association, 3225 N. First Avenue, Tucson, Arizona 85719)

That's STELLAR!

No Instruments Required!

A Tahitian navigator named Tupaia helped English explorer Captain James Cook navigate when he sailed from Tahiti in 1768. To the captain's amazement, Tupaia guided the boat to the small island of Rurutu, using nothing but the stars for navigation. No matter where the ship lay on the sea, the wise navigator could always point to the direction of Tahiti.

Time Travel

For the book *Contact*, the author, Carl Sagan, consulted a relativity expert to figure out how the main character could travel to distant stars, and return to find that no time had passed on Earth. Many believe that the expert replied that the character should travel through a "wormhole," a shortcut through space and time. Some scientists are examining if wormholes actually exist.

It's Constant!

Hubble's law, named after the famous astronomer Edwin Hubble (1889–1953), says that the farther a galaxy is from us, the faster it's moving away. This makes it plain as day: The universe is expanding.

Legends of the Stars

In Homer's epic poem, *The Odyssey*, Odysseus sails toward home from Kalypso's island, guided by the stars . . .

*Glorious Odysseus, happy with the wind, spread sails
And taking his sea artfully with the steering oar who held her
On her course, nor did sleep ever descend on his eyelids
As he kept his eye on the Pleiades and the late-setting Bootes,
And the Bear, to whom men give also the name of the wagon,
Who turns about in a fixed place and looks at Orion
And she alone is never plunged in the wash of the Ocean.
For so Kalypso, bright among goddesses, had told him
to make his way over the sea, keeping the Bear on his left hand.*

Universal Jokes

How many astronomers does it take to change a light bulb?

None. Astronomers prefer the dark.

What do you call a film about bugs in space?

A sci-fly movie.

The renowned Professor Bignumska, lecturing on the future of the universe, had just stated that in about a billion years, according to her calculations, the Earth would fall into the Sun in a fiery death. In the back of the auditorium a tremulous voice piped up:

"Excuse me, professor, but h-h-how long did you say it would be?"

Professor Bignumska calmly replied, "About a billion years."

A sigh of relief was heard. "Whew! For a minute there, I thought you'd said a million years."

—*a famous joke, as told by Douglas R. Hofstadter*

Machos and WIMPs?

What's out there in the universe? Some scientists think that cold dark matter makes up most of the universe, either in the form of black dwarfs or brown dwarfs. But others believe that WIMPs (weakly interacting massive particles) and MACHOs (massive compact halo objects) make up most of the universe. But don't expect to find them any time soon: This "stuff" is undetectable!

Create the Cosmos

What's out there? A lot. In fact, scientists are still puzzling about the matter that makes up our "known" universe. There is still much to learn. Take a "wall" in your classroom and create the universe on it.

Break up into groups of two and choose a celestial object from one of the following: our solar system, the Sun, neutron stars, pulsars, variable stars, binary stars, star clusters, superclusters, and galaxies. Then spend a week gathering information from the Internet and the library on your subject.

Based on what you know about the motion of stars and your object, try to determine:

What is it made of?
What is its role in the universe?
What is its age?
Where is it located?
Is it changing?
How far away is it? Can we see it with the naked eye?

Pin your data and pictures on a designated wall or bulletin board in the classroom.

Now try something a little closer to home. The star closest to us is the Sun. The movement of the earth around the Sun has a great effect on our world, as each year goes through a cycle of four seasons. For thousands of years, humankind has lived according to a yearly cycle, for planting and harvesting crops. How has the constant and predictable movement of the Earth around the Sun influenced human civilizations through time?

Divide into two groups once again. One group can study the northern hemisphere, and the other can study the southern hemisphere. Find out about the history and culture of the different regions in your area, and create a detailed farming calendar, indicating which crops are planted and harvested, and at what time of year. When the research is complete, both groups can compare their calendars and discuss why crops follow different cycles in different areas of the world.

Ready for the ultimate challenge? Enter this or any other science project in the Discovery Young Scientist Challenge. Visit *discoveryschool.com/dysc* to find out how.

ANSWERS

Gravity Wells, page 13

1) It collects in the center of the bowl because of gravity.
2) The sand outside the bowl represents interstellar dust.
3) The larger the bowl the more sand is collected. This represents a more massive star.

Powers of Ten, page 15

1) 1.2×10^{12} m
2) 9.5×10^{15} m
3) 4.0×10^{16} m
4) 7.7×10^{20} m
5) 2.1×10^{21} m
6) 1.4×10^{26} m

Solve-It-Yourself Mystery, pages 26–27

Sonny's idea was the only scenario that meets Mr. Cosmo's assignment. In his story, the Orion IV is heading towards our solar system, and the "small yellow star" is actually the Sun. (Note that Mr. Cosmo's opening sentence didn't actually say that the Orion was taking off from Earth.) Our solar system has a belt of asteroids between Mars and Jupiter, and a ring of comets at the outer boundary. Ray's idea won't work because despite the name, a planetary nebula isn't likely to contain planets. A red giant is a star at the end of its life, and so it is unlikely to provide the energy necessary to support life on any orbiting planets. Celeste's idea is inaccurate for two reasons: First, as a cloud of dust and gas, a nebula has no distinct center of gravity, and so it can't support a planetary system. Also, the star at the center is a pulsar, evidence that the star's supernova explosion has already taken place. So the Orion IV is not in danger of being destroyed by a supernova, simply because the explosion has already happened.